Climate Change and Me!

i

How We Help Control, And Combat Climate Change. To Change the World.

By: William J. Harp

Table of Contents

Chapter

10……… Solution

Chapter One: Acknowledgement

Climate change! It is a scary thought. A global extinction. Just imagine everyone you have ever known were dead. Including yourself. And imagine the whole planet lifeless. That thought keeps me up at night. That is something nobody wants.   And yet there are people who deny this. They are called the climate deniers. They say it is not real. It is made up. I cannot understand them when you can clearly see by looking at the planet that climate change is happening.

Usually the climate deniers are older white men that are going bald. But if you explained to them that climate change was causing their hair loss, well I believe that they would stop climate change completely. But since that is not going to happen, the fear is real, and that gets me upset, and disgusted, like someone kicked me in the stomach. But then I realize that the people who deny climate change probably never finished high school and did not go to college.

But we do not live in a world where it is that simple. I must keep telling myself to get the word out, explain to all people that I run into on the street, that climate change is real, and it's a threat on our lives, Most people already believe in climate change, but the ones that don't well I must convince them with passion and facts that climate change is real.

I believe that everyone can get the word out about climate change and it would change the world for the better. We just must be passionate and persistent. Just imagine that the people you told about climate change, and then those people telling more, and those people getting the word out. It would spread across the globe. If everyone took a personal commitment to stopping climate change. it would change the world.

I tell people what happens if you have a pet fish, and one day you neglect it. Nothing happens. Then the next day you neglect it again, and still nothing changes. But over several days or a week the fish dies. Because the temperature in the water changes, the water becomes more toxic, and dirty, and the fish cannot sustain the harsh conditions. Or the delicate flower that cannot receive enough water to survive. then eventually it wilts and dies.

That is the same with climate change. and with us knowing that nothing will change after one day, then the next. then one day it is over, our environment will change, and we will not be able to live. We will not be able to grow food. The temperature will be too extreme. and poof were gone. Or the delicate flower that cannot receive enough water to survive. then eventually it wilts and dies.

The science has supported climate change, and the main reason for climate change is people. A single human being causes the most damage. But get billions of people together and that spells the end. There are 7 billion people on this planet. A single person will have the most catastrophic effect on our climate throughout a person's lifetime with all the choices that person makes. Whether we drive a vehicle or ride a bicycle. Our personal decisions. Down to the air we breathe.

Climate change is affected by all the fossil fuels that are used. Someone using the computer or game station or watching Netflix, all of that requires power. And that comes from the power plant, which generates millions of tons of greenhouse gas emissions. It is hard to think about. The damage that is causes on our planet. And it is even harder to try and prevent climate change when people do not agree. But together we are stronger, and we can accomplish anything together. If we were a million strong or 2 million strong or more, the power behind that would do so much good for the world. Being able to get things accomplished. Everyone agreeing to stop this atrocity. And have a world in which we can live in, without fear of becoming extinct.

America is one of the leading countries contributing to climate change which is killing the earth. Our planet can only take so much before it reaches the breaking point.

Chapter2: WHAT IS CLIMATE CHANGE:

First, we must understand climate change to fix climate change. It is when the atmosphere captures co2 carbon and is unable to release it into space. And there in our atmosphere the carbon builds up creating a greenhouse effect, essentially raising the temperature several degrees. When this happens, it affects every living thing on earth. This also means that the plant life would not be able to keep up with the overabundance of carbon co2 to breathe in and convert it into oxygen. So, then we are left with an atmosphere full of co2 carbon instead of oxygen. Because of more carbon put into our air, it pushes out the oxygen. It will cover our entire planet. Our scientists are concerned about climate change because our planet is warming up faster than in any other time in our history. The average temperature is not average anymore. Its higher. Just look at the ice caps that are melting. Even the weather patterns have changed. It is everything that is not renewable. From the air we breathe out, to the products we use.

The causes of climate change are everything we do that is not renewable. CO2 is the contributing factor. But humans! We are the biggest contributor to climate change because we are the main producers of CO2. The air we breathe out is CO2. All the products that we use, from air conditioners and appliances, and anything that has an engine on it, uses fossil fuel (gasoline). Factories cause an increase in our CO2 production, especially oil companies, and the cars that people drive. The overabundance of electricity that we use. Burning fossil fuel. The use of cell phones and computers, and the lights in our homes. The games on

PlayStation that is played. The night light in my bedroom. It is the chemicals the farmers use on their crops, that wash into our water system and get released into the air. It is the jet fuel from planes, it is the waste in the city dump. It is our bathroom waste. It is everything that we do, and we must be more aware of what we do, has an impact on our planet. and that effects everyone. What I do, here in my home, effects people on the other side of the planet.

Mankind is the cause of climate change. Harsh, I know. But it cannot be sugar coated; it cannot be swept under the rug. It must be understood that humans are to blame.

Since the industrial revolution, the world's factories have put out more pollution and carbon CO_2 at an alarming rate. We have the plastic companies, the oil and gas companies, the steel mills, the lumber yards. We have the waste we put in the sewer in my town and yours, and the garbage that is produced. None of this disappears. It must go somewhere. Then it comes down to the individual person causing climate change. The air we breathe is CO_2 carbon. The waste a single person causes has a tremendous effect on our climate and the world. We are the biggest polluters on the planet. That candy bar wrapper that someone threw on the ground, or the propane tank that the grill uses, causes climate change. One just works faster than the other. Even though these companies and corporations cause the bulk of climate change, the individual runs these companies and it cannot be overlooked. This is just too serious to ignore.

Chapter 3:   THE EFFECTS of TEMPERATURE:

Let us look at the first effect. The temperature increases.

First, with the raised temperature it will evaporate more water, which in time will completely disappear. And with no water, no life. We need water to live and so do the plant life and animals. That leads to the extinction of all life. But there are oceans full of water someone might say. And yes, they would be correct, but it is not a drinking water source and since there is only a limited supply of drinking water, that would evaporate before the oceans. So, our drinking water supply would basically vanish. With the increase in temperature it creates the snowball effect. It prolongs hotter months which would increase our use of air conditioners in our cars, homes, and offices and the use of more water for our food supply. including plants and livestock. The hotter weather would create a migration of insects to move to a more ideal place. which could mean a shortage of certain vegetation and crops. All of this happens globally. The displacement of insects could destroy crops across the world since those insects have a positive and negative effect on crops.

OCEAN EFFECTS:

The water temperature increases as well, and this leads to the migration of sea life to look for more suitable waters. The fishermen would be left with nothing to fish for. You have the melting of the polar ice caps, which can cause flooding and tidal waves. The added pressure of water displacement could cause more earthquakes, and tsunamis. Then there is the change in the water direction of the currents like the Australian current featured in Finding Nemo which affects our weather patterns. Then the water evaporation would cause severe storms and down pours all around the world causing flooding. Also, there are the coral reefs with millions of sea life around them also the coral reef too is alive. Which could be exposed above sea level, displacing that sea life. Or it would cause the coral reefs to disappear.

When the fish are in a different climate, that too can bring an extinction level to our sea life. No more fish.

The polar ice is diminishing and causing a disruptive sea flow and displacement of all sea life habitats.

AGRICUTURAL EFFECTS:

In the extreme weather conditions that it creates, droughts and flooding and wildfires are a result. It affects the farmers growing their crops.

They must have a healthy soil to be able to grow and nourish the world. But without water that help the crops grow, there will be no food.

There are also insects that have migrated because of higher temperatures and that has affected the natural eco system from which these insects survive. Some insects are beneficial, and some are not, depending on the living conditions that influence crops.

To combat these bad pests that have migrated, farmers have taken up insecticides to combat the insects from attacking their crops. And the use of these chemicals has negatively affected their soil. The runoff of these chemicals has infiltrated our drinking water. All these actions of one thing affecting another, then another, have consequences for everyone it even effects our health.

You cannot have healthy people if you do not have healthy water or crops.

Without a healthy soil to feed our plants and livestock and crops, you cannot have healthy people. These insecticides cause pollution and illness and when it is mixed in with the soil, the plants and crops absorb those chemicals. Also, when it is mixed in the drinking water, our livestock consumes that water, infecting them with the chemicals. Then when the people consume those vegetables and meat, we are infecting ourselves with those same chemicals, which cause cancer and other terminally ill diseases.

Then the chemicals get evaporated into the air and then becomes part of the rain somewhere else across the globe. It is a life-threatening cycle.

There are 3 things that affect food production

There is a high concentration of CO_2, and the CO_2 molecules are what captures the sun's radiation and sends it back to earth. Also, 1) CO_2 is what is used for plants to produce food, through the process of photosynthesis, and when more CO_2 is produced, it causes the plants to produce less nutrients. Because of the higher CO_2, the plants are less able to take up the nitrogen from the soil. 2). Another issue is increasing heat. It is harder for the plants to grow because it decreases the time allowed to grow and they mature faster. and then plants lose water more rapidly because of the evaporation from the plants due to higher temperatures. 3). It also raises the stress level on plants since they will not get enough cooling hours. All of this leads to what all cattle eat or do not eat and whether they get enough nutrients to survive. The smallest negative effect in the soil affects everyone. The shortage of being able to grow crops, the less food is produced. Then there would be a food shortage.

Need to cut greenhouse gas emissions which negatively effects our wheat, hay, corn, rice, all crops.

WEATHER EFFECT:

Sometimes It is scary to hear thunder and see lighting, and the rain pounding on the house in the middle of the night during a thunderstorm. But we will see weather so extreme that it is hard to imagine. Droughts would consume a whole city. The town you live in could eventually be a desert. No water, no plant life, just a vast amount of dried up cracked earth like Death Valley. and endless days and nights of hot 100 + degree weather that would last for 4 or 5 even 6 months. This would cause the increase of more resources to combat the extreme weather and in turn cause more CO_2 in our atmosphere making the problems worse

.   Then the massive evaporation of water would cause severe downpour and thunderstorms like buckets of water being dumped over a town which would cause flooding and water damage to everything. Even the plant life could not sustain overabundance of water. It would drown them. Like watering the flowers in your mom and dad's garden, too much water would kill the flowers, vegetables, and grass. The resources needed to repair the damage to homes and buildings caused by the flooding. Would put a strain on wood and the manufacturing companies that make more plastics and steel. Production from those companies would put even more carbon CO_2 in the air. And the cost to replace and repair would cause economic disaster as well. Since there is a limited supply of materials. The poorer neighborhoods would not be able to afford to rebuild or repair. Leaving millions homeless. Worldwide.

FOOD SUPPLY EFFECT:

With no resources to provide water for animals and plants. there would be no food, only what our scientist can engineer in a lab: Genetically modified foods. But they would not be able to produce enough for billions of people. Approximately 7 billion people are on this planet. Essentially there would be no food to eat. it is called famine. And everyone will experience it. When our bodies are malnourished. then we are susceptible to disease.

Our food supply would be attacked by migrating insects trying to survive themselves. The food supply would be nonexistent.

MIGRATION EFFECT:

The migration of people is due to climate change. Extreme temperatures would cause displacements of whole towns. Extreme weather is causing an influx of migrants worldwide in Europe and America. It puts even more pressure on communities with less money, creating overcrowding and more human waste, disease, and illness along with it. Limited opportunities for jobs would become less with more people living in those communities. This really affects everyone.

MENTAL AND PHYSICAL EFFECT:

The mental and physical effect would be dangerous to every person.

The lack of food and water would deplete the vitamins and nutrients needed for your body. We would basically starve. Our physical health would deteriorate and increase in disease, and malnutrition would take over. Then the mental aspect would cause depression and anger and sorrow. I believe no one would be mentally equipped to handle this type of catastrophe, fighting over what food is available, what housing is left. But in the end, there would be no food, no life, on this planet.

SOCIAL EFFECT:

Climate change causes social injustice around the globe.

What kind of society would we have? People already fight over the smallest of things, like road rage and someone bumping into someone. And kids making fun of other kids. Playground fights are something so trivial that it is forgotten the next day. The discipline the teachers give to their students for disrupting the class, the argument you had with your friend or with your parents or brother and sister are social issues that we deal with daily. They seem pale in comparison, to the devastation and destruction of our planet because of climate change. All these issues pale in comparison to the global extinction that is climate change. Climate change does not care what color your skin is. It does not care if you are overweight. And if we can put aside these issues and focus on what really matters, our society can become better.

ANIMAL EFFECT:

Look at our cattle the increase in temperature would cause the grass they feed on to dry up, leaving them with little or no options for food consumption. Then there is little access to what water is left. And that would kill off our cattle. The cattle would migrate to better fields to try and feed on until eventually there are none left. The cheeseburgers people eat or the toppings for pizza would be gone even the cheese and milk and butter cows produce. But with our cattle gone and our plant life shriveled up and dead then what is left for people to eat? Nothing. Then human beings would starve and become extinct. This is no exaggeration It has been proven by scientists around the world.

ECONOMICS EFFECT:

Everyone should pay a climate tax. There should be a digital currency. The amount of energy it takes to make paper money and coins has too much of a negative effect on our environment. The implementation of a digital currency would eliminate the negative effects and help combat climate change.

If only our government could do something to change this, to combat it and stop it from happening. Pass a law that stop companies from producing CO_2 or tax people for using CO_2 or tax the companies that make the straws or cups through a carbon tax. They were able to pass a law to stop using trans fats. Why can't they pass a law to stop climate change? It does not seem too difficult. But here we are in a time of crisis and nothing is being done. If they know what causes climate change, why we cannot stop it.

If a town in any country is too poor and not well funded or equipped to deal with a natural disaster like flooding because of climate change then those people cannot afford to rebuild that town or replace the damaged property. Then what happens is the migration of those people to another place. It is unfair that this affects people in poverty the most. We owe it to society to take responsibility for everyone, so this does not happen.

DISEASE EFFECT:

Just imagine if you were fishing in a pond near your neighborhood and you catch a couple of fish and bring them home to eat them. You cannot because of all the chemical run off due to everyone using fertilizer sprays on their yards and the farmers using them on their crops. Then the evaporation of those liquid chemicals turns into a gas and is absorbed into the atmosphere and travels across the world. Then when it rains that chemical breakdown is combined with the rain and is dispersed worldwide, which affects all animal life, especially humans. It takes those chemicals that drain off the yards directly into the pond, so it is absorbed through the fish and they incur the negative effects of those chemicals like cancer and other illnesses, we consume the fish and animals which would make us sick with cancer and other diseases.

And those same chemicals affect the growth of crops.

The accumulation of animal and human waste in a concentrated area is increased through migration if the population doubles in a town. That waste must go somewhere, and the resources that town has is further burdened. The concentration of more people would cause the flu or some other communicable disease measles or tuberculosis to Spread faster. The effects would be crippling.

Because of climate change there has been an increase in negative effects in our immune systems with an increase in allergies and other chronic diseases and illness.

With an increase in consumption and with diseases and overgrowth of population, we need to eat and consume less. It is with supply and demand that our society is based on. We should instead look at it through the supply side, instead of the demand aspect of It, and understand that eating more is not better. If we consume less, there are less chemicals used in agriculture that would positively effect climate. We must change our diet to help climate change. For example, there is 1 gram of protein in a lentil plant vs 1 gram of protein in a cow, but the cow produces more methane gas than the lentil plant. It would be beneficial to eat the lentil than the beef. But with the consumption of beef we eat 5 times more than the recommendation from the U.N., and we eat twice as much as needed per the FDA. And we waste about half of that. Just eat less and do not waste food. Remember also there are starving children around the world. The lucky ones get to eat once a day which is usually just a small bowl of rice. It is hard to think about it when we are stuffing our faces with McDonalds, and we see people just throw half a hamburger away but knowing that half could help someone else who has nothing. I imagined that if McDonald's and other restaurants would cut their portion size and donate the difference to starving children, that would help dramatically.

What I eat affects me individually, locally, and globally. Eating more diverse foods puts less of a strain on our soil, our production companies and transportation. And its benefit mother earth most of all.

Chapter: 4 ISSUES THAT CAUSE CLIMATE CHANGE:

Let us go down a small list of issues causing climate change.

Humans:

Animals:

Corporations:

Automobiles:

Every engine that uses gasoline:

Power plants:

Coal:

Guns:

Just to name a few.

Humans are the biggest cause of climate change and contributes to the largest amount of greenhouse gas emissions. The calculations are based on what we create as a species plus what we use. And the carbon that we breathe out daily is just too much carbon dioxide. Another fact is humans are the only animal on the planet, not waste positive. The choices that every individual makes on a minute by minute decision, whether that person thinks about it or not, will affect how much carbon they will contribute to the world. So, we must be aware of what we are doing. Not only does its affect me individually, but it affects everyone in the globally.

Using guns and firearms are dangerous for the environment. Every time a bullet is fired it releases carbon into the atmosphere. Guns throughout history have caused climate change and killed millions of people. Not only do guns help contribute to climate change, but they are dangerous to humanity. In a society where we discuss our conflicts worldwide at the United Nations, we have no need for guns in a civilized society. Guns should be banned for climate change and for being unnecessary. I believe our society is at a point where we do not need guns to live in peace and harmony. It is truly the goal to get rid of weapons that cause damage to the environment and can kill people. We certainly do not need them. We can come together to banish guns from the face of the earth.

Animal life causes CO2 or the methane that they expel, which comes from all animals. The bigger the animal the more carbon emissions they are producing. This is just a small example of animal life which includes cows, horses, dogs, cats, bears, lions, insects, birds, fish, and other species. Unfortunately, they are not able to stop or reverse climate change. And that leaves the decisions with us on how to appropriately handle how we proceed on correcting it whether new laws are implemented on the farmer to control more of the greenhouse emissions by cleaning up after the animals or by an extended tax or by limiting the amount of animals a farmer can have. The solutions are limited by your imagination.

Corporations.

The decisions a corporation makes whether it is making a product or selling a product will affect climate change. Making a product depends on where it is being made and the laws that allow how much carbon can be released in the air. What product they are making, plastics or glass or whatever, will have a devastating effect on our planet. And if they are selling a product then, the means of how they sell that product will affect climate change whether they use television ads, or they use billboards or flyers or newspapers. It will cause issues with our climate. These corporations use office products that affect our planet right down to the paper clip they use or the pencil or what type of computer. It is amazing to think that just the littlest things we do influences our climate. The amount of electricity that these occupied buildings use is staggering, so they should be outfitted with solar panels.

Automobiles

Everything about an automobile causes severe carbon emissions right down to the plastics and the gasoline (fossil fuel) that it uses. This is one of the most contributing factors in climate change. Automobiles are the second contributing factor in climate change right below humans.

Power plants and coal

Power plants use coal to produce electricity, and that coal is mined from the earth. It has taken millions of years to produce and it just simply cannot be replaced. More importantly, when that coal is burned, it releases carbon into the atmosphere, doing so much damage that the earth may never recover. Coal is one of the main things that produces electricity. That must change. It is also a limited substance.

Circle of influence starts with our plate. Your friends may try something new since you are eating something new.

Chapter 5:   CO2 EFFECTS:

The constant buildup of CO_2 cannot escape out of our atmosphere, like a congested freeway. so, it just stays there and once there is enough CO_2 trapped in our atmosphere the sun's rays will shine down on our planet, and they will be trapped in the carbon in the air which will hold the sun's rays. Then it will continue to build, and the heat and temperature will rise causing catastrophic global warming.

For starters, as the temperature rises the heat will dry up the soil. And with the increase in temperature the evaporation of our lakes, streams, and oceans will take effect. And when the water evaporates into the air, then it will create violent storms.

In a wealthy community with good infrastructure and insurance it will disrupt their lives temporarily, but if a violent storm happens in a place that is poorer and doesn't have a good infrastructure or the people don't have insurance or adequate housing, then that same storm can cause devastating effects that will last for years and more importantly would cause death to those people. More tsunamis and tornados and earthquakes. And flooding. Would come possibly consuming entire towns. This will be a see-saw effect: blistering heat for months on end then violent storms for months. Then one day when it decides to become 200 degrees Fahrenheit, it will just turn the planet into one big desert. The water that provides life for our plants will diminish, and the plant life will start to dry out and die. And we will be left with no vegetation or crops to grow food to feed people. The soil will dry out making it harder for plants. Then the insects will disappear because of no plant life like bees and birds who pollinate the flowers so they can in turn produce. All will disappear.

Once our sea level starts to decrease and eventually it will destroy our sea life, down to the microorganism. And the ocean water will decrease and eventually evaporate. Our sea life will be wiped out completely. And our water will be gone.

Will have no beautiful whales or dolphins to look at, and no water to fish in. It would be a vast planet of emptiness. There would be no oceans left. Or life.

Right now, our water supply is being poisoned with all the chemicals that are being used by the lawn companies that spray for weeds and the pesticides that are used either by your parents or a company comes by to spray for insects. or for the lawn. All those chemicals get washed out through the rain and washed down the sewer drain for the city. From there it flows out into the local rivers, and from there it goes to the ocean.

Those chemicals are cancer-causing products, and they are poisoning our water supply. When there is enough chemical wash off into all the rivers and oceans and when that water evaporates and comes down as rain, it would be like acid rain burning through everything.

Our drinking water will start to disappear and our food. There would be no grass or hay for livestock like cows and pigs so they could eat. Once they go extinct then we will have no vegetables or meat to eat. Then the humans will die off. Or worse, if you can survive, these extreme conditions, I imagine it would be like living in a house with a temperature of 200 degrees or even more, like living on the planet Venus.

Could anyone sleep when It is that hot without passing out from heat exhaustion. Imagine living in your bedroom at night lying in bed and it is as hot as a sauna and sweat pouring down your face. I do not think I can handle that. Or anyone else. It would drive someone mad.

Another scenario plays out that with the abundance of C02 in the atmosphere that will overtake the oxygen element and cause humans to lose air by squeezing out the oxygen because too much carbon is in the atmosphere. And then humankind would all suffocate to death.

Look at this through a pie chart if our atmosphere is at 60% oxygen and the other 40% is a mixture of other elements and CO2 is one of them. Then what happens if more companies produce CO2 and it becomes 60% or our atmosphere and oxygen is only 40% or less? If you have something that grows and grows bigger then eventually it will take over and push the other element out.

No matter how this is played out. every possible scenario ends with the extinction of human and animal and plant life. It would literally kill everything on the planet.

Chapter 6:   CONTRIBUTORS:

What else contributes to climate change besides everything that's non-recyclable. It is also oil companies, corporations. But the 2nd biggest contributor are the cars and planes. It is the plastic that is produced like straws and cups and all the trash that goes to the city dump which then produces CO_2. It is the natural gas that most homes use. It is the aerosol sprays, it is the school bus, it is power plants that produce electricity; it is construction sites.

Humans produce the most CO_2 gas. It is an endless list. Everything you do contributes to climate change. The pencil you are using. How was it made? Was it environmentally friendly? What about the paper you are using? I believe that now in this time that all schoolwork should be done on a computer and reduce the destruction of trees for paper. We must weigh the pros and cons. Is the carbon from a computer worse than using a pencil and paper?

There are mass migration issues because of climate change, and millions of people are being displaced in their own country. They would have to seek refuge somewhere else because their towns cannot sustain them to live. And this leads to overpopulation and with that the food supply will be overburdened, medical facilities will be over filled, and so will the education system. It will be harder to have adequate housing and more difficult to discard waste and trash, and it eventually becomes unsanitary. And disease will start to spread killing millions of people.

The possibilities of a vaccine for everyone that would eliminate a lot of illness. And quite possible a vaccine in the future that could limit the output of CO_2 gas from humans, and animals.

The effects of poverty are another cause of climate change. Since low income people are stuck in a system that is hurtful, they are forced to receive the cheapest products or none, and they are not environmentally friendly. They do not have the financial means necessary to properly deal with climate change. They do not have the insurance to rebuild if a disaster destroyed their home or the money to relocate to a better environment. Most people live without any savings; they live on a day to day basis. And the migration issues should be proof enough to show displacement from their home country where they are no longer able to farm their land due to climate change.

We cannot hide from climate change, and we need to take responsibility.

We are only here because our planet allows it.

Chapter 7:   MYSELF

      Climate change and me. Well I thought about this for a long time and I kept saying to myself that I am just one person. What can I do to help combat climate change? What significance could I bring to the table? After all, I am just one person, and some say one person cannot make a difference. But I realized that I could make a difference, even if it is just me. I know that it is a huge task and that for myself I would have to take baby steps as a single person to help stop this. And just like any problem someone deals with, first you must know what you are up against. Maybe you could join a climate organization or non-profit. Like any problem we must find out what is causing it, to correct it.

Chapter 8:   BEHAVOIR:

What can I do to stop climate change when it starts with me? Stop using straws and plastic bags. Make the conscious decision to do what it takes to stop climate change. Try to take public transportation when possible or use a bicycle. Do not overeat. When going out to eat a meal, split it with a friend or family member because wasting food is contributing to climate change. Do not forget, it also causes obesity. Eat smaller portions. No one likes to be overweight. Recycle. You can re-use a water bottle or lunch pail. Just sit down and write a list of what you can change in your daily life. When brushing your teeth, instead of leaving the faucet on, turn it off until you need to rinse your toothbrush. When taking a hot shower, the amount of carbon used to heat that water is astounding unless you could convince your parents to get a solar water heater. There are many things we can do to make this planet better.

You can talk to everyone about climate change and how it effects our earth, our food system, our water, our agriculture, and our health.

Our planet will not sustain life. The land will stop producing crops. And the sea water will evaporate. There will be nothing left but a barren waste land like mars.

Our food will not be able to grow due to the extreme weather changes. It will be impossible to grow food or raise cattle. It will be an entire worldwide famine/starvation like what we see in third world countries. What will happen globally will affect us locally.

Our water supply is being poisoned by the chemicals and eventually will be undrinkable. Then once climate change is in full swing our water will evaporate. No water equals no life....

Our health will be affected. Because of all the chemicals used for vegetation and for cattle it will result in drastic health issues such as cancer. and other deadly diseases. And the proof is already here with the increase of more and more people getting allergies. That is just the beginning. there will be more deadly viruses. and diseases.

Talk with your brother and sister about climate change and its effects on everything. Talk with your parents about it. Explain that we are not comfortable with some of their decisions and that changes must be made. Talk with your friends and let them know how serious this is. Talk with your classmates and teachers and try to organize an event at school. Talk with the principle, maybe organize an earth day. Take to the streets with a school drive or a signature petition to send to congress to enact laws to combat climate change. Make the conscious decision of not having children. Hold politicians accountable locally and nationally. We cannot isolate ourselves from climate change, it is a global problem, and since it is global, it breaks down to politicians and corporations and individuals to tackle this problem. It must be a global solution.

Make more connections with neighbors, farmers, your doctor, your classmates, teachers. Reach out to people around the world through social media, such as Instagram and Facebook. A petition on Facebook could generate billions of signatures and commitments to help with climate change.

Individually to make a change is to vote. Be adamant with your parents about voting and that who they choose will take the responsibility of correcting these issues.

Make personal choices that benefit a green world. Our personal choices will ultimately decide how this planet will survive.

Change the environment. Movement will offer millions of people the collaboration of coming together to make a better world, not only physically for the planet but socially as well and economically.

There is a word that should be used with creating climate change and that is ethics. One can correct their bad behavior or from behaving unethically towards our planet. And this should be pointed out every time it is witnessed.

Environmental sustainability is something everyone should strive for. It is the end game, and we need to control what is produced and used to control our environment.

Climate change is so serious that I am crying. My heart aches and my palms are sweating. It is very disappointing and nerve wracking and stressful.

Chapter 9: THE CURE:

Climate change seems just too big for me to make an impact. I am just one person. What could I possibly do to help combat climate change? Well, climate change and me. It starts with one person and builds on that to reach millions of people. So that one person can and should be you. You are important, your thoughts and actions count. One person can make a difference.

Talk with your parents. Let them know how you feel about climate change, how it affects you emotionally and physically. Maybe your parents never took the time to understand climate change. Let them know what it is and what its effects are. Let them know that it is personal and that you feel responsible for it. And that you are serious about taking responsibility to help stop climate change.

There are several things you can do. When you get older and decide to buy a car, maybe you buy an electric car, or if you live in a big city, you take public transportation. Or you can ride a bicycle to work or put solar panels on the roof. But for now, we can start to grow our own vegetables and fruits at home in the back yard or on the roof top. That would stop some carbon footprints from the fossil fuel burning trucks, and planes delivering vegetables to the store. Or you could choose a green career.

But for now, there are important things we can do. We could conserve energy by unplugging electronic devices when not in use. For instance, if the devices that have a remote control have that red light on it, it is still using electricity. If you have a power strip or power surge protector, you can just plug in the devices to that and push the button to turn all those devices off instead of having to unplug each one separately.

It is our environment. We need to learn how it works

We can plant a tree to prevent more CO_2 carbon from entering the air.

We can turn the thermostat in the house up during hotter months and lower in colder months.

We can use fans when it is hot to save on electricity and wear coats or sweater when its colder to save that way. That would be better for the planet

We can carpool or ride our bicycle. If the family must go to the store, we can make it an adventure with everyone riding their bikes and each family member carrying a recyclable bag home. We can use a reusable cup instead of the plastic bottle water and coke cans. And drink tap water.

We can recycle all our trash like the aluminum cans and glass and plastics, and if the sanitation dept does not recycle in our area, we can write letters to the local leaders to get them to start. But also, there are stations that accept recyclable goods that your parents can take too.

Buy products that use environmentally friendly material.

We can conserve the water we use. For instance, when brushing your teeth, do you leave the faucet running or do you turn it off after you run your toothbrush under the water? We can turn the water off. we can pay attention to how long we take showers and conserve water by being more efficient at taking showers. Do you take a 30-minute shower or a 10-minute shower? How long is too long? We can get our parents to replace toilets with water saver and energy saving appliances.

We can use less paper towels when cleaning and use washable rags.

Your parents can get an electric lawn mower or a push mower with no engine to cut down on carbon output.

We can shop at environmentally friendly stores.

We can think of ways to leave a less aggressive climate footprint. We can buy biodegradable products and buy local.

We can talk to school officials about conserving energy and going solar.

Next time they pass a school bond, include solar equipment.

Try to get school busses to go electric.

Maybe if people were in concentrated area's they would be more likely to use public transportation. And a better use of public utilities since it would not have to be on such a wide electric grid.

Cafeterias should be revamped on improving food because so many classmates just throw away trays full of food. And it should be discussed how to improve and eliminate all that waste.

You can start by telling family members about the seriousness of climate change and that each one of us needs to be more aware and to help combat it. Then you need to talk to your friends and let them know. And if they do not take it serious then maybe they are not a true friend.

Then you need to talk to your classmates and teachers to have them get involved by writing letters to congress and local leaders or the U.N. by expressing your concerns and possibly offering solutions that nobody has thought of. Maybe have a school strike for climate change on earth day or offer a suggestion to congress to enact a tax on everyone to pay their fair share of combating climate change. Or do a fundraiser to help fund climate change relief.

Also, we have a choice to make on whether we have children. Since a person is the greatest threat to climate change, having children is going to be a choice all of us make to better keep our planet clean. But the possibility is that when climate conditions get worse, congress might enact laws that prohibit having children or limiting the number of children a person can have. But we are trying to save the planet before it gets that bad.

Our lives and future generations live depend on it.

Chapter 10: SOLUTION:

I am just a kid. What do I know about climate change? I am worried about school and friends/ All I know is what I see. Attitudes must change. We all must care about our fellow human beings. We are all trying to survive on a planet together.

We are here to save the world one person at a time.

Climate catastrophe, economics, health, social anarchy, over population lead to more illness and disease, waste plus garbage. people fighting over scraps. Utter chaos'.

Everything you buy has to do with climate change, consider those companies that you buy from. Are they helping to fight climate change, or are they contributing to climate change? Everything we do in our lives affects climate change.

Corporations like McDonalds, electric companies, clothing co. basically any business can charge a climate tax.

Scientists are already making genetically modified foods in preparations for food shortage from climate change.

Everything you do affects climate change. What you wear or what you eat or how you get around your city all contribute to climate change. We need serious people to solve this crisis. And that starts with you.

Everyone needs to be educated on climate change. Our education system should implement a curriculum to teach students the effects and impact of climate change.

For people with pre-existing conditions that are uncurable there should be help with assisted suicide for people who choose such an option.

We should also do away with parenting and have the community raise the children. Not all parents exercise the best parental controls.

The only way to really combat climate change is to change human behavior, to perceive that our lifestyle contributes to climate change, to somehow live in a society that has sustainability that does not derive from carbon.

But until we each take responsibility the next best thing is to institute a climate tax.

Going through a drive thru restaurant would have a climate tax to use the convenience of keeping that vehicle running to pick up your food

Or a grocery store would implement a climate tax on using plastic bags.

There would be a climate tax for bullets since that causes carbon emissions

There would be a climate tax for cable television

There would be a climate tax on gasoline to have the privilege of driving.

There would be a climate tax at all restaurants

There would be a climate tax through the electric companies

There would be a climate tax through the toll roads

There would be a climate tax through the payrolls of corporations

There would be a climate tax through the mail service

There would be a climate tax on all cell phones.

These are just a few ideas of what we can do to help stop climate change

But just imagine if there was just a one-dollar climate tax added to every cell phone bill. That would generate $200,000,000. Two hundred million dollars a month in the united states alone! And if it went worldwide then there are 500,000,000,000. Five billion cell phone users and this amount of revenue would generate every month. That is a lot of money, and if that was combined with the other ideas, then that would be a nice start to stopping climate change.

Not only would that help stop climate change, but with extra funds it could stop poverty. Give equal equity to everyone, a standard of living, and end famine worldwide. It could also help with social justice education and our friends in the LGBTQ community. It could help fight racism by providing financial equity. It also could end borders. A world without borders with everyone living in peace. With climate change stopped and world hunger gone, and poverty ended, this could be a world in which we would have a true paradise. Everyone would be equal. A one world government where every possible aspect of a standard of living is given to everyone. So, no one would have to worry about going to bed hungry or not being able to have the necessary clothing or roof over their heads. To have a world heath care which everyone is treated. To end paper currency and convert to a digital currency. And by eliminating the production of the paper and coin currency would cut down on carbon emissions. This would lead us all to be a global citizen.

I am ready to fight climate change, are you?